THE THAMES

OXFORD TO LONDON

TWENTY ETCHED PLATES

BY

DAVID LAW

"Slow let us trace the vale of Thames." *Thomson.*

18

British Library Cataloguing-in-Publication Data
A catalogue record for this book is available from the
British Library

A Short Introduction to the History of the River Thames

The River Thames takes its name from the Middle English *Temese,* which is derived from the Celtic name for river. Originating at the Thames Head in Gloucestershire, it is the longest river in England, flowing a total length of 236 miles, out through the Thames Estuary and in to the North Sea. On its journey to open water it passes through the country's capital, London, where it is deep enough to be navigable for ships, thus allowing the city to become a major international trade port.

The earliest evidence of human habitation on the river is a Neolithic bowl (3300-2700 BC), found in the river at Buckinghamshire. Other Pre-Roman sites have also revealed watermills, navigations, burial mounds, and settlements along its banks. As with many geographical features of Europe, it was the Roman Empire that realised its strategic and economic importance. In the first century BC, under Emperor Claudius, they built many fortifications along the Thames Valley. They also established a trading centre at the river's lowest point on two hills, now known as Cornhill and Ludgate Hill. They built a bridge there and named the settlement Londinium, a settlement that would eventually become the City of London.

During the Middle Ages the river attracted great prosperity, being a hive of activity in the fishing, milling, and pottery industries. However, the success brought with it unwanted attention, and in 870 AD the Vikings swept up on the tide creating havoc and destroying buildings such as Chertsey Abbey. When William the Conqueror arrived in

1066 AD he was rightly concerned with protecting the Thames Valley. He built many castles along its banks, such as those at Windsor, Rochester, and the magnificent Tower of London. This cemented his strategic position and gave him a base from which to control the rest of the country. He was also responsible for giving us the Domesday Book (1086), a hugely detailed survey, which gives us great insight into the activities on the Thames in the Middle Ages.

In the 16th and 17th centuries the river became a hugely important tool for transporting goods such as timber, livestock, and foodstuffs, from Oxford to the capital. It helped the City of London grow with the expansion of world trade and the wharves became packed with shipping vessels. It was also during this period that the country endured a series of cold winters that froze the Thames. So solid was the sheet of ice that 1607 saw the first Frost Fair, where tents were set up on the river offering various amusements such as ice bowling. By the 18th century, London was the centre of the mercantile British Empire and the river became one of the busiest waterways in the world. Docks were expanded and locks were built, allowing for easier navigation and even greater method of trade.

Among the many uses of the Thames, unfortunately one of them was as a dumping site for the city's waste products. This had been the case since the Middle Ages, but the more populous the banks of the river became, the more it became an environmental hazard. This problem reached its zenith with the 'Great Stink' of 1858, when the stench from the raw sewage in the river caused the abandonment of a sitting at the House of Commons. A concerted effort was then made to clean up the river and the construction of massive sanitary sewers was undertaken. This was part of a great engineering drive in the wake of the Industrial Revolution that saw pollution decline, water sources become cleaner, and railway bridges reducing the congestion on the river.

Although the Port of London remains one of the UK's main ports, the growth of road transport has largely superseded the Thames as a medium of transport. There are also far fewer heavy industries utilising the river, and as a resulted it is cleaner than it has been in hundreds of years. The Thames has played a key role in the development of a nation. Its story, both geographically and with its human uses, is a long and fascinating one. As John Burns, MP for Battersea, said in 1929 'The Thames is Liquid History'.

PREFACE.

THE Thames valley includes a great variety of different types of that peaceful scenery which is accustomed to be called "English." Take only a few examples. At Hart's Hill, at the Quarry Woods, and at Cliveden, there are ranges of hills whose wooded sides rise steeply from the banks of the stream, whilst at Medmenham, at Bisham, and at Bourne End, the hills recede from the river, and form a distant background to the landscape of meadows and cornfields. Then there are park-like glades of stately timber at Nuneham and opposite Datchet; rustic mills at Iffley and at Mapledurham; rushing weirs at Pangbourne and at Hambledon; quiet and shady backwaters at Cleeve, Sonning, and Hurley; whilst here and there the "willowy isle" diversifies the course of the stream; and the tower of many a village church peeps over the trees which surround it. Nor should it be forgotten that there are such world-renowned prospects as the distant views of Windsor and of Oxford. Small wonder, then, that the banks of the Thames should become the home of the artist, and the scene of many a holiday trip.

It is hoped that this book will recall many pleasant memories to those who are familiar with "The silver winding Thames," and that it may induce others to become acquainted with its many beauties thus described by Thomas Love Peacock :—

"The plain, where herds unnumber'd rove,
The laurell'd path, the beechen grove,
The lonely oak's expansive pride,
The spire through distant trees descried,
The cot with woodbine wreathed around,
The field with waving corn embrown'd,
The fall that turns the frequent mill,
The seat that crowns the woodland hill,
The sculptured arch, the regal dome,
The fisher's willow-mantled home,
The classic temple, flower entwined,
In quick succession charm the mind,
Till, where thy widening current glides
To mingle with the turbid tides,
Thy spacious breast displays unfurl'd
The ensigns of the assembled world."

OWEN J. DULLEA.

October, 1881.

CONTENTS.

The ornamental head and tail-pieces are facsimile reproductions of pen-and-ink sketches by the artist.

OXFORD.

THE Folly Bridge is certainly not the most interesting of starting places for the journey down the Thames. A few strokes, however, bring the boat alongside of the barges used as boat-houses by the various College clubs, of which a sketch is given above. They are pleasantly sheltered under the trees of Christ Church Meadows, whilst the high ground about Kennington forms a distant background to the meadows on the right bank.

Where the river makes its first bend at the "Gut"—the scene of many a hard struggle in the College boat races—is a favourable spot for a general view of the University city.

> "Now, rising o'er the level plain,
> 'Mid academic groves enshrined,
> The Gothic tower, the Grecian fane,
> Ascend, in solemn state combined."

The spire of St. Aldate's on the left, is succeeded by Sir Christopher Wren's domed tower over the entrance to

Christ Church; then follow the spires of the Cathedral and of St. Mary, the University Church; the classic tower of All Saints, designed by Dean Aldrich, architect, logician, and musician; the twin towers of All Souls; the dome of the Radcliffe Library, and the massive tower of Merton. Last, and perhaps most beautiful, on the extreme right, are the pinnacles of Magdalen, of which there is a reminiscence below.

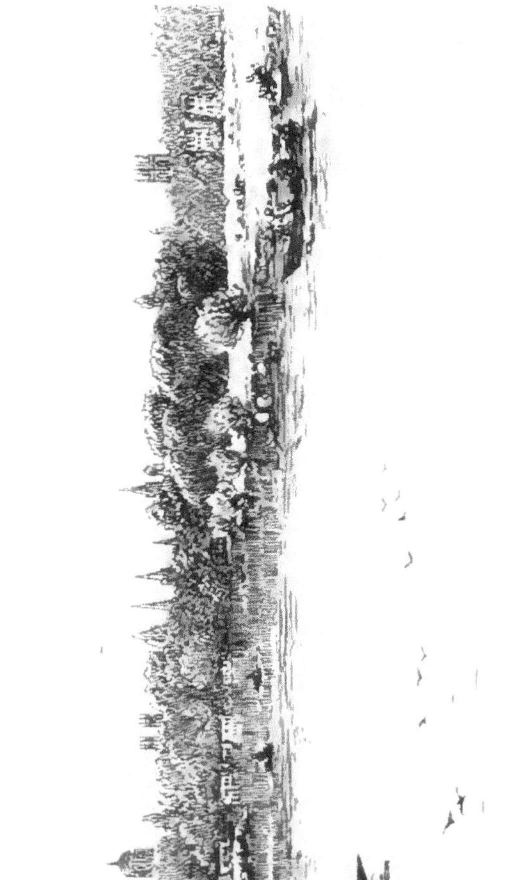

IFFLEY MILL.

AFTER leaving Oxford, Iffley Lock, the first stage in the journey down stream, is soon approached, and

> "Sinking o'er the level mead,
> The classic domes and spires recede."

Just above the lock, the embattled tower of the old Norman church, the goal of many an archæological pilgrimage, peeps invitingly over the trees. Once through the lock-gates, a good view is obtained of the well-known mill, whilst holding up the boat against the stream. The red-tiled roof and whitened walls of the mill-house contrast pleasantly with the green foliage by which it is surrounded. Beneath the tarred out-buildings, works

> "The dark round of the dripping wheel."

A little foot-bridge over the weir-paddles connects the mill with the lock island on the left hand, where waves a group of the tall poplars which form frequent landmarks in Thames scenery.

Between the Kennington Railway Bridge and Rose Island, a parting glimpse is caught of the spires of Oxford, and the pinnacled tower of Magdalen stands out prominently in this last view of the University city. After Sandford Lock and Mill—what a contrast, with its tall factory shaft, to that left behind at Iffley—the stream continues its course between level banks until the "tufted woods and sloping glades" of Nuneham Courtney rise in front. To these soon succeed the island and rustic bridge, the favourite haunt of Oxford picnic parties, a sketch of which is given on the previous page.

ABINGDON.

THE spire of Abingdon is seen in the distance, soon after the railway-bridge which terminates the woods of Nuneham. Below the Lock, however, it disappears from the view, and but little is seen of the town until the boat has passed the bridge—of which a sketch is given above—and the group of islands, on one of which is built the "Nag's Head" Inn. At the bend in the stream is the little foot-bridge, seen on the left of the etching, under which the Ock flows to join the Thames. It rises in the range of the White Horse, which, as Drayton sings,

"Sent presents to the Thame by Ock, her only flood."

The almshouses—by which St. Helen's churchyard is, cloister-like, surrounded, and whose chimney-stacks cluster round the spire in the etching—are very interesting. The most ancient are those at right angles to the quay. On the gable fronting the river, there is a curious painting of the old market cross destroyed by the Roundheads in the Civil Wars. Those almshouses parallel to the river date from the reign of George I., and are a curious example of an endeavour to combine the features of a cloister with early Georgian architecture.

The church itself has a very noble interior. Although its Gothic is not of the best period, and it is somewhat wanting in length, yet its height, and the double aisles on both sides of the nave, impart a dignity which the details fail to supply.

GORING.

TIRED with the long and rather uninteresting reach below Wallingford, the eye rests with pleasure on the change after Moulsford Railway Bridge. The banks once more become wooded, and the unpretending village church at Moulsford, surmounted by a small bell-cot, is seen close to the water's edge. Cleeve Lock is then near at hand. The reach between this and the next lock, although one of the shortest, is perhaps one of the most beautiful on the river. In front, there is a delightful view of the broad expanse of the Streatley hills. But the chief beauty of the reach lies in the series of small weirs and well-wooded islands by which the stream is broken up. There are many quiet backwaters whose placid surface is overshadowed by wide-spreading trees,

> "Where the duck dabbles 'mid the rustling sedge,
> And feeding pike starts from the water's edge,
> Or the swan stirs the reeds, his neck and bill
> Wetting, that drip upon the water still."

GORING.

To the left of Goring Lock is the narrow stream fringed by dwarf trees, which feeds the mill shown in the etching. Above it rises the roof and tower of the village church. The interior, consisting of a chancel and lofty nave, with one aisle, is disfigured by a screen and some other modern additions executed in the most approved carpenter's or churchwarden's Gothic. It, however, contains some brasses and other points of interest, and the tower has some excellent Norman features.

STREATLEY.

ON the opposite side of the stream to Goring, and connected with it by a long straggling bridge with a toll-house in the centre, is the sister village of Streatley. Looking down the backwater towards the Berkshire shore, from about the same spot whence the view of Goring Lock is taken, the "Swan" Inn occupies a prominent position in front of the spectator. The gabled roof, the walls covered with climbing roses and Virginia creeper, and the rustic porch entwined with honeysuckle, combine to render it one of the most charming resting-places on the river. On the right hand, embosomed amidst the luxuriant foliage of some noble timber, is the square and embattled tower of the church.

The background to the scene is formed by the great range of the Berkshire downs, here known as the Streatley Hills. There are hills in many other portions of the Thames valley: at Nuneham, below Marlow, and at Cliveden, they rise abruptly from the water's edge, and are thickly clothed with rich masses of foliage. Again, at Basildon, at Fawley, and at Medmenham, their slopes form a distant background to the meadows and corn-fields. But in no other part of the stream, do the hills rise so boldly, and with such a bluff outline as at Streatley, and it is this characteristic which constitutes the distinctive charm of the scenery in the adjoining reaches.

PANGBOURNE WEIR.

AFTER Goring, the hills alternate on either side of the river with a charming variety. First, on the right bank, the Streatley Downs rise boldly from the water's edge; then a bend in the stream leaves red-bricked Basildon House behind, and the hills on the left draw near. Here the slopes and cliffs are thickly covered by the beech trees of Hart's Wood, and some reedy islands mark the position of a former weir. Meanwhile, the hills left behind at Basildon have formed a delightful background to the broad expanse of fields and meadows which stretches between them and the arc-like course of the river. They now again approach the stream, and, as Pangbourne draws in sight, upraise a chalky barrier which hides the railway, here running parallel to the water. The turfy slopes leave but just enough space for the road and towing-path.

A stroll in the village of Pangbourne is always pleasant. Its two streets—if they may be dignified by such a name—wind about in the most serpentine fashion. Here and there the little Pang is to be seen dancing merrily over its pebbly bed. Across the bridge, of which a reminiscence is given below, is the sister village of Whitchurch with its mill and wooden-spired church.

After such a saunter, what is more pleasant than to rest on one of the felled trees in the timber-yard by the side of the weir bay, and to watch the water rushing by in swirling eddies. In the background are the trees in the grounds of Combe Lodge on the further side of the river, whilst on the left is that delight of anglers, the old "Swan" Inn, covered with creepers; a sketch of it is given on the previous page.

MAPLEDURHAM.

LEAVING Pangbourne and Whitchurch Bridge behind, a ridge of gently ascending hills runs parallel to the stream on the left hand. Their verdant slopes are here and there diversified with the foliage of shady clumps. Then a bend in the river brings in sight one of the many well-wooded islets with which the Thames is studded, and to the left are seen the gables and chimneys of Hardwicke House, which has several associations with Charles I. Soon after, the narrow entrance to Mapledurham Lock, shaded by a friendly row of trees, appears on the right; whilst, looking towards the opposite bank, is seen the subject of the etching.

Mapledurham is deservedly one of the most renowned spots on the Thames. The wooden sides and moss-grown roofs of the mill, over which is seen the tower of the church, fully realize the ideal of the poet and the painter, to say nothing of the scenery by which the spot is surrounded. If the views above the lock are delightful, those below it have another variety of charm, and are even yet more beautiful. The sinuous course of the stream is split up by several islands, and the channels between them afford glimpses of the rushing weir. On the right bank, with their boughs overhanging the surface of the water, are the noble avenues of Purley Park,

> "Where waving groves a chequered scene display,
> And part admit, and part exclude the day."

On the Oxfordshire bank are seen the dormers and clustered chimney stacks of the Elizabethan manor house of Mapledurham, one of the few English mansions which can boast of remaining in the possession of the same family for the last three centuries. Its owners, the Blounts, are descended from a Picardy family, one of whose scions settled at Mapledurham in 1489. His grandson, Sir Michael Blount, was Lieutenant of the Tower under Queen Elizabeth, and to him the present house owes its erection about 1585. One of his descendants was the father of "the fair-haired Martha and Teresa brown," the well-known friends of Pope.

SONNING.

AFTER passing through Caversham Lock, the tall chimney shafts of Reading stand out, unpleasantly prominent, against the sky. The river winds so much that they are not left behind until

> "Clear Kennet overtakes
> Her Lord the stately Thames, which that great flood again
> With many signs of joy doth kindly entertain."

After this, the stream still wanders for some time through broad and open meadow land, as if uncertain which way to take. But a change in the character of the scenery is at hand. The river approaches a belt of ample trees, whose foliage soon affords a delightful overhanging shelter to the towing-path, and amidst a leafy wealth Sonning lock appears in sight.

The locks up the river are frequently rendered pleasant resting-places by the gardens cultivated by the lock-keepers in their leisure moments. The garden at Sonning is quite

Virgilian with its roses and bees, and the lock-keeper has, in his cultivation of the poetic Muse, a further resemblance to Virgil's pastoral swains.

Sonning is best seen, as in the etching, from the opposite bank below the bridge. To the left is the village street, and at its corner the "White Hart" Inn, the pleasant garden of which slopes down to the water. In the centre of the view, connecting the village with the mill island and the opposite shore, is the solid old brick bridge, whose low arches seem scarcely high enough to allow a passage for the boat. Over the trees of the churchyard rises the grey tower of the church, itself in its turn surmounted by the grand chestnuts of Holme Park.

HENLEY.

BELOW Sonning, as above it, the river flows through a succession of verdant meadows, which form rich pasture for the many grazing cattle. The stream pursues its course with many windings until Shiplake is reached. Here, the church, picturesquely perched on the side of a hill, overlooks "where Thames amongst the wanton valleys strays," and at this point the river is joined by "the Loddon slow with verdant alders crowned."

At Wargrave, the tower of the church, as shown in the sketch above, is seen over the fine elms which surround the churchyard. It boasts of very good bells, and the effect of a muffled peal across the meadows, which separate the village from the church, is very impressive.

After the chalk cliffs and woods of Park Place, the well-known landmark, the tower of Henley Church, soon comes in sight. It is an excellent type of the square embattled tower common to the Thames valley. The church itself is a good specimen of late Perpendicular, and has several interesting monuments, notably that of Lady Periam, who is represented in all the pride of ample Jacobean ruff, stomacher, &c. The wife of Sir Godfrey Kneller and the master-builder of St. Paul's Cathedral are also buried here. Perhaps the most interesting memorial is that to the French general, Dumouriez, the victor of Jemappes. Averse to the excesses of the revolutionary party, he became a proscribed fugitive, and sought safety in several countries successively. He at length found a retreat in this little riverside town, where he ended his days.

MEDMENHAM ABBEY.

BELOW Hambleden lock, the hills on the right bank of the stream bear some resemblance in their bold and rolling character to the Streatley Downs. Perched half-way up their slope is the red-bricked mansion, Culham Court, whilst at the foot of the range is the cluster of little islands known as Magpie Eyot. The hills then quit the bank, and the river winds round a tongue or promontory of open meadow land. Close to the shore of the bay thus formed in the opposite bank is Medmenham Abbey. The remains of the abbey have really but little antiquity. Such fragments as there are, have in fact greater associations in connection with the riotous fraternity of Bubb Doddington, the poet Churchill, and John Wilkes, whose motto "Fay ce que voudras" is still to be seen over a doorway, than with the austere Cistercians by whom the abbey was founded. It is the charm of the scene itself which constitutes the attraction of Medmenham. The ruins are partly covered by a luxuriant growth of ivy, whilst such of the walls as are exposed to sight are plentifully bedecked by lichens, and have that delightful grey and yellow tint belonging to age. Then a group of fine trees clusters around, and a lovely background is supplied by the slopes of a distant ridge of the Chiltern Hills.

BISHAM.

WITH many windings amongst reedy islets, the stream brings the boat beneath the cliffs of Harleyford. Here and there the chalk peeps out from amidst the thickly mantling foliage. On the opposite bank is that most enticing of villages, Hurley. Amongst its attractions are Lady Place with its reminiscences of the Revolution of 1688, the old church with the remains of the monastery, and the old-fashioned rustic inn.

Returning to the river, the towing-path is carried across the lock cuttings by frequent and simple wooden bridges, one of which is shown on the next page. Then there are the shady backwaters and straggling weirs—

"Where bordering hazle overhangs the streams,
Whose rolling current winding round and round,
With frequent falls makes all the woods resound."

Soon after Temple lock and mills, the spire of Marlow Church is seen in the distance over the meadows to the left of the stream. To the right is the tower of Bisham Church, almost shrouded from sight amongst a group of tall trees, and on looking back, as in the etching, the view includes the Abbey, in which few perceptible alterations have been made, and which still retains its grey and old-world aspect. It was founded by one of the great Montacutes, Earls of Salisbury, and became the burial-place of the family. Here was buried the Thomas, Earl of Salisbury, "the mirror of all martial men," who was killed at Orleans, and of whom Shakespeare says:—

> "In thirteen battles Salisbury o'ercame:
> Henry the fifth he first trained to the wars;
> Whilst any trump did sound, or drum struck up,
> His sword did ne'er leave striking in the field."

Then the bones of the great King-maker, and his brother, the Marquis of Montacute, were laid here after the fatal field of Barnet, and one of the last who found a resting-place within its walls was the unfortunate son of the Duke of Clarence, who was imprisoned and beheaded by the Tudors. But their ashes were all scattered at the Dissolution, when the Abbey was assigned by the great plunderer to his cast-off Queen, Anne of Cleves. It then came into the possession of the Hoby family, whose grand monuments are to be seen in the adjoining church.

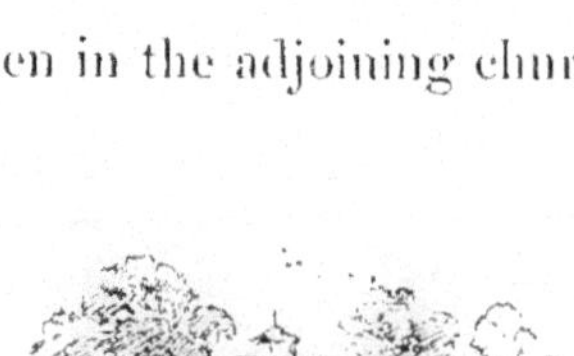

D. LAW

MARLOW.

THE reaches above and below Marlow Bridge are in the midst of some of the most charming scenery on the river. In the words of Thomas Love Peacock,

> "Delight shall check the expanded sail
> In woody Marlow's winding vale."

Just above the bridge is a favourite resting place for house-boats, which find many pleasant nooks amongst the trees and bushes on the right bank, as shown in the sketch above. But it is by the tribe of anglers that Marlow is most frequented. It forms a "happy hunting ground," where

> "The patient fisher takes his silent stand,
> Intent, his angle trembling in his hand:
> With looks unmoved, he hopes the scaly breed,
> And eyes the dancing cork and bending reed."

After the suspension bridge and lock, the rush of water from the two mills quickly carries the boat down the stream.

To the right is the entrance to the back-water leading to the weir, of which a sketch is given below. It is a lake-like expanse of shallow water, and from it is obtained a good view of the high ground, clothed with the Quarry Woods, which stretches away to Winter Hill on the left, and which forms a charming background to the scene throughout the reach below Temple Lock. Slightly lower down the main stream, the towing-path crosses the little wooden bridge shown on the right of the etching. Looking back at this point, the lock gates, the gables and roofs of the mills group well together. Above them rises gracefully the spire of Marlow Church, now at a sufficient distance to hide the feeble crockets and details which were considered correct Gothic "when George the Fourth was king."

COOKHAM.

AT Spade Oak Ferry, there is a sudden and entire change in the character of the scenery. In the reach just past, the hills rise in an abrupt ridge from the river's bank; their sides are thickly clothed with wood down to the edge of the water, which is overhung by the branches of the trees, and a series of well-covered eyots divides the stream into two channels. But after the ferry and the boat-builder's yard, broad and level meadows take the place of the hills, which here form a distant limit to the river valley, and the course of the stream is wide and open.

Below the wooden railway-bridge at Bourne End, the river makes a broad sweep, bay-like, into the pleasant pasture-land on the right. In the distance, the trees hide from sight the village of Cookham, whilst over the foliage is seen

> "The old church tower
> In majesty presiding o'er the vale
> And all her dwellings."

COOKHAM.

This view forms the subject of the etching, whilst the sketch on the previous page, looking back towards the railway-bridge, gives the reverse of the picture.

Cookham can certainly be numbered amongst those places which do not fall short of their reputation. The village street, the old-fashioned inn, "Bel and the Dragon," and the open green, recall forcibly the pictures of Frederick Walker, whose brush, during his frequent sojourns here, was inspired by the quiet and homely charm which surrounds the village. He now rests peacefully in the churchyard. Again, the view from the bridge, looking upon the river with the tower and glorious beech woods of Hedsor in the background, forms a prospect which has few equals in the scenery of the Thames.

CLIVEDEN.

ON leaving Cookham Bridge, the course of the river is rather perplexing to the uninitiated traveller. At this point, the stream diverges into four different channels. To the right, two channels lead to Cookham Mill, and to the backwaters round Formosa Island. The stream to the extreme left flows to Odney Weir and Lord Boston's eel-bucks. Next to it is the course to be followed by the boat, a long narrow stretch of water leading to the lock.

Immediately below the lock gates, the view is very striking. On the left bank, the wooded declivities are so steep that, with the lofty trees on the opposite bank, the river seems as if it were flowing through a narrow mountain gorge, whose sides tower imposingly on either hand. Then come the ferry and summer houses, which are soon succeeded by three islands amply fringed with reeds and flags. Just below this point the glory of Cliveden culminates. The

Buckinghamshire banks of the stream, which is here widened into a broad expanse, are shaded by the overhanging boughs. In many places the branches almost touch the surface of the water. In thickly serried masses the trees rise one above another,

> "The tall oak his spreading arms entwines,
> And with the beech a mutual shade combines."

Surmounting the leafy heights in the distance is the façade of the mansion. Although the present house is modern, there is much interest connected with the former associations of the spot. Beginning with the Duke of Buckingham, who

> "In the course of one revolving moon,
> Was chymist, fiddler, statesman, and buffoon,"

it has numbered amongst its owners Lord Orkney, one of Marlborough's lieutenants in Flanders; Frederick, Prince of Wales; and latterly the ducal families of Sutherland and Westminster.

BRAY CHURCH.

THE woods of Cliveden are quickly succeeded by Boulter's Lock and by Maidenhead Bridge, of which a sketch is given above. Then comes an expanse of open country, and over it is seen the row of fine poplars which form a landmark for some distance above and below Bray.

The two most interesting features in the village of Bray are the almshouses and the church. The Vicar of Bray, renowned in song, has alas! been proved by modern research to be a most mythical personage. The red-bricked Jacobean almshouses, or Jesus Hospital, are redolent of bygone days. The trimly kept, old-fashioned garden, with its trees cut into fantastic shapes, and the quadrangle with the ivy-covered chapel and chaplain's house, carry the imagination far away from the nineteenth century. One of the approaches to the churchyard is very picturesque. It is formed by a low gateway underneath an old timbered cottage with an external staircase. According to an inscription on one of the timbers, the cottage dates from the year 1448. The greater part of the noble church itself was probably erected about 1300, and the fine oak roof to the nave and aisles is of this period. The tower, which is seen in the etching over the trees in the vicarage garden, was probably erected about a century later.

Within the precincts of the churchyard, but entirely distinct from the church, with which it is coeval, is a chantry dedicated to the Virgin. In 1683, it was converted into the school-house, and was used as such for more than one hundred and fifty years. Within its walls, Hearne, the well-known antiquary, was educated.

WINDSOR CASTLE.

THE swiftly-flowing current in the reach between Monkey Island and Water Oakley quickly brings the boat to the first distant glimpse of Windsor Castle. After Surley Hall and its poplar trees, comes Boveney Lock. But, as Drayton sings, there are still many windings in the stream before a closer view is obtained of Windsor,

> "Whose most delightful face, when once the river sees,
> Which shews herself attired in tall and stately trees,
> He in such earnest love with amorous gesture woos,
> That looking still at her his way was like to lose;
> And wandering in and out, so wildly seems to go,
> As headlong he himself into her lap would throw."

There is perhaps none of the ancient edifices of England which occupies a position at once so beautiful and commanding as Windsor Castle. Viewed from the Park, the eye wanders amongst sylvan glades, or revels in the distant perspective of avenues of majestic growth, at length to rest on the distant terraces and towers of the Castle. Seen from the river, as in the etching, the Castle uprears an imposing mass against the sky. From what is known as "George the Fourth's" Tower on the left, there is a continuous line of battlements to St. George's Chapel and the Curfew Tower, with its steep French-like roof, on the right. The mighty mass seems to dominate over the town, whose roofs cluster confidingly around its base.

Denham has expressed this union of beauty and strength in one of those quaint conceits dear to authors of his time:—

> "Windsor the next (where Mars with Venus dwells,
> Beauty with strength) above the valley swells
> Into my eye, and doth itself present
> With such an easy and unforced ascent,
> That no stupendous precipice denies
> Access, no horror turns away our eyes:
> But such a rise as doth at once invite
> A pleasure, and a reverence from the sight."

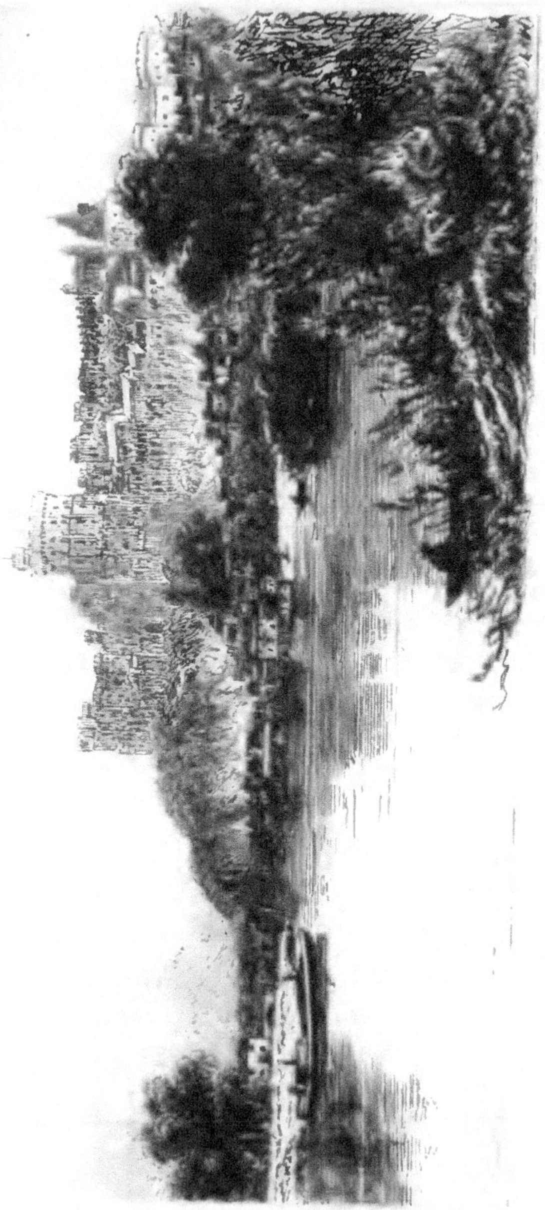

ETON COLLEGE.

LEAVING the Berkshire banks of the river where

"Majestic Windsor lifts his princely brow,"

there is another link with the past on the opposite shore,

> "Where Eton is at hand to nurse that learned brood,
> To keep the Muses still near to this princely flood;
> That nothing there may want, to beautify that seat,
> With every pleasure stored."

The best general view of Eton is obtained from Romney Island, and forms the subject of the etching. To the right are the trees which surround the playing fields on the opposite bank, whilst the green sward of the Fellows' Eyot intervenes between the river and the buildings forming the east side of the cloisters. Over their roofs appear the turrets and pinnacles of the chapel, of which Henry VI. laid the first stone in 1441. In his emulation of the noble foundations of William of Wykeham at Winchester and Oxford, he intended that the present structure, grand as its proportions are, should form only the choir of his magnificent collegiate church. To it, he would have added a nave and aisles, and, judging from the massive buttresses, there is little doubt that the building would have been vaulted. But his intentions were never carried out: the troublous latter years of his reign prevented the fulfilment of his cherished designs; and it is only in the present century that an endeavour has been made to render the interior more worthy of the exterior.

HAMPTON COURT.

THERE are two aspects of Hampton Court, which recall two distinct series of historic associations: the Tudor and the Dutch. Viewing the Palace, as in the etching, from the opposite bank of the river, where it is joined by the

"Sullen Mole, that runneth underneath,"

or passing through the Entrance and Clock Tower Courts, the many turrets and clustered chimneys, the battlemented walls and oriel windows, the Della Robbia busts of the Cæsars and the magnificent Great Hall, with its elaborately carved roof, recall the last of the English prince-prelates, of whose erection it was said,

"The kynges court
Should have the excellence,
But Hampton Court
Hath the pre-eminence."

Here the latter years of the pageant-loving Henry VIII. were chiefly spent. Within its walls, Somerset and Northumberland struggled for the protectorship of the young king, Edward VI. Here Mary and the gloomy Philip II. passed their honeymoon, and Elizabeth, who, like her father, loved gaiety, held her festivities.

The Dutch associations of the Palace commence as soon as the Fountain Court is reached. This, and the whole of the eastern front of the building were designed by Sir Christopher Wren for William of Orange. Here, as Macaulay says, in his splendid banishment from his native country, he found some consolation in erecting another Loo on the banks of the Thames. In the intervals between his foreign campaigns, Hampton Court and Kensington Palace formed his favourite residences; and it was in Bushey Park that his horse stumbled over the fatal mole-hill.

RICHMOND.

COMING down the river from the last lock at Teddington, an open reach displays its straight course as soon as Twickenham Eyot is passed. On the one side, the broad expanse of stream is bordered by the trees in the Twickenham meadows: on the opposite bank are the stately groves which infold the mysterious and deserted-looking Jacobean mansion, Ham House. Amongst its treasures, the Van Dycks and Lelys alone would be sufficient to supply several country houses with Caroline ancestors, whilst its Caxtons have caused an enthusiastic old bibliophile to call it a "wonderful book-paradise." In the distance, the stream is divided by the little Petersham Eyot, above which

> "Fair groves and villas glittering bright,
> Arise on Richmond's beauteous height."

This is one view of Richmond from the Thames. Another —the subject of the etching—is obtained from below the

Bridge, erected about 1770, by means of a tontine, one of the forms of the craze for lotteries in the last century which survived longest. The best stand-point for this view is Cholmondeley Walk, a pleasant riverside footpath, which occupies the frontage of the ground where formerly stood the three palaces successively erected at Richmond. The first palace was a favourite resort of the early Edwards and of Richard II. The last-named monarch, whose first queen, Anne of Bohemia, died here, took so violent a dislike to the place that he caused it to be pulled down. The second palace owed its erection to Henry V., but it did not outlast the century, for it was burnt down in 1498. On its ruins immediately arose the magnificent structure of Henry VII., which became a favourite residence of the Tudor sovereigns. Within its walls were entertained Philip I. of Spain, the Emperor Charles V., the sullen Philip II., and Eric IV. of Sweden, one of the many suitors of Elizabeth, who herself died under its roof. The Commonwealth, and the reigns of Charles II. and Queen Anne, saw the gradual dismantlement of the palace, until at the present day the only visible remnant of its former grandeur is the old red-brick gateway facing the Green.

WESTMINSTER.

THE last stage in the journey from Oxford to London is now reached. In the words of Drayton,

> "Then Westminster the next great Thames doth entertain,
> That vaunts her palace large, and her most sumptuous fane:
> The land's tribunal seat that challengeth for hers
> The crowning of our kings, their famous sepulchres."

Few parts of London have undergone more change during the present century than the banks of the river at Westminster. Lambeth Palace and the grey Lollards Tower still stand unchanged, but the Albert Embankment and the long range of detached buildings which forms St. Thomas' Hospital have taken the place of the wharves and barges which formerly lined the Surrey shore. The Westminster Bridge built in 1750 by the Swiss architect Labelye, has been replaced by a structure which, whatever its merits, is certainly not seen to advantage at low water. On the

Middlesex shore, the Victoria Embankment with its plane trees stretches across the bay which formerly exposed a hideous expanse of mud at low tide. But the greatest change is where the vast Gothic pile of the elder Pugin and Sir Charles Barry, flanked by the massive Victoria and Clock Towers, has taken the place of the old Houses of Parliament.

It was on the view from Westminster Bridge that Wordsworth, at the commencement of the present century, wrote the sonnet beginning,

"Earth has not anything to show more fair;"

and no more appropriate parting with the series of beauties illustrated in this book, can be found than in the words with which he was inspired by the river in his "In Memoriam" for the poet Collins:—

"Glide gently, thus for ever glide,
O Thames! that other Bards may see
As lovely visions by thy side
As now, fair river! come to me.
O glide, fair stream! for ever so,
Thy quiet soul on all bestowing,
Till all our minds for ever flow,
As thy deep waters now are flowing."

Made in the USA
Monee, IL
06 November 2026

68010594R00042